Making of America Project

Proportions of pins used in bridges

Making of America Project

Proportions of pins used in bridges

ISBN/EAN: 9783337156800

Printed in Europe, USA, Canada, Australia, Japan

Cover: Foto ©berggeist007 / pixelio.de

More available books at **www.hansebooks.com**

PROPORTIONS

OF

PINS USED IN BRIDGES.

BY

CHARLES BENDER, C. E.,

MEMBER OF THE AMERICAN SOCIETY OF CIVIL ENGINEERS, AND
OF THE GERMAN SOCIETY OF ENGINEERS IN BERLIN.

NEW YORK:

D. VAN NOSTRAND, PUBLISHER,

23 MURRAY AND 27 WARREN STREET.

1873.

PROPORTIONS

OF

Pins Used in Bridges.

A portion of this work without the
mathematical deductions was presented to
the American Society of Civil Engineers,
February 19th, 1873.

It is not the least of the merits of skeleton structures, such as are built by the best
American constructors, that any of the parts
can be exactly calculated. The determination of strains acting in pins is not excepted
from this statement, though it is connected
with the application of some of the finer
laws of the theory of elasticity.

All rules referring to the size of pins can

be immediately deduced from the few physical principles on which the theory of elasticity is based, and without new experiments. On the contrary, sound practice would have further advanced and mistakes been avoided, if, at least, a scientific examination of the subject had been undertaken previous to experiment, in order to arrive at a distinct opinion about what confirmation by test was required of the theory of elasticity. There are three questions to be examined in the present paper, viz. :

1. What is the law of distribution of the pressure caused by the tie bar on the bearing surface of the pin.

2. What is the law of distribution of the shearing strain over the cross-section of the pin, and

.3. What is the value of its bending moment in any of its sections.

The solution of the first actually includes also that of the third, but it is more convenient to treat the two separately.

For this purpose, consider a top chord pin of a truss bridge, represented by Fig. 1.

The originally straight pin M M is resting on bored bearing surfaces N N, of a top chord casting, and carries two flat eye bars of the width " b " and the thickness " t " placed outside of the casting. This is the arrangement of any pinned bottom chord, or the chains of a suspension bridge, where the outermost bars produce strains analogous to those borne by the parts here shown.

Under strain, the pin M M will be pressed in to the bearings N N, which are supposed to be of a length B C$=l$, and of the depth d. Both bearing surfaces E F, between pin and bars, will be compressed, and a greater share of the total pressure absorbed by the parts of the pin closest to the face of the casting. The pressure per square unit at F will be greater than at E, and likewise the pressure at B than at C. The pin is supposed to accurately fit into its bearings, and (for the present) friction between the cylindrical bearing surfaces is disregarded.

So far as the half cylindrical bearing surface or half circle of any cross section of the pin is concerned, the pressure will not be any more uniformly distributed than is the case for the different points of lines E F and B C. The pressure of a pin with play first will be concentrated at a point, but under the load a set will take place, and the pressure be more uniformly distributed over the diameter $r\,r$. The law of this distribution is very complex, and will not be followed up here. It is sufficient for the present to consider the pressure as being uniformly distributed over the diameter of the pin, which, however, is somewhat less than is actually the case. The diameters of pins found under this supposition will be a little *too small;* reasons will be given for assuming this basis of the calculation.

It may be added that the strain per square inch will not only decrease in a single ratio with the increase of the pin, but that a larger pin will cause a more uniform distribution of the pressure; the importance of the *least practicable play* in the pin hole is evident.

The pressure per square inch is in direct proportion to the amount of depression; consequently the problem is the same as to find the

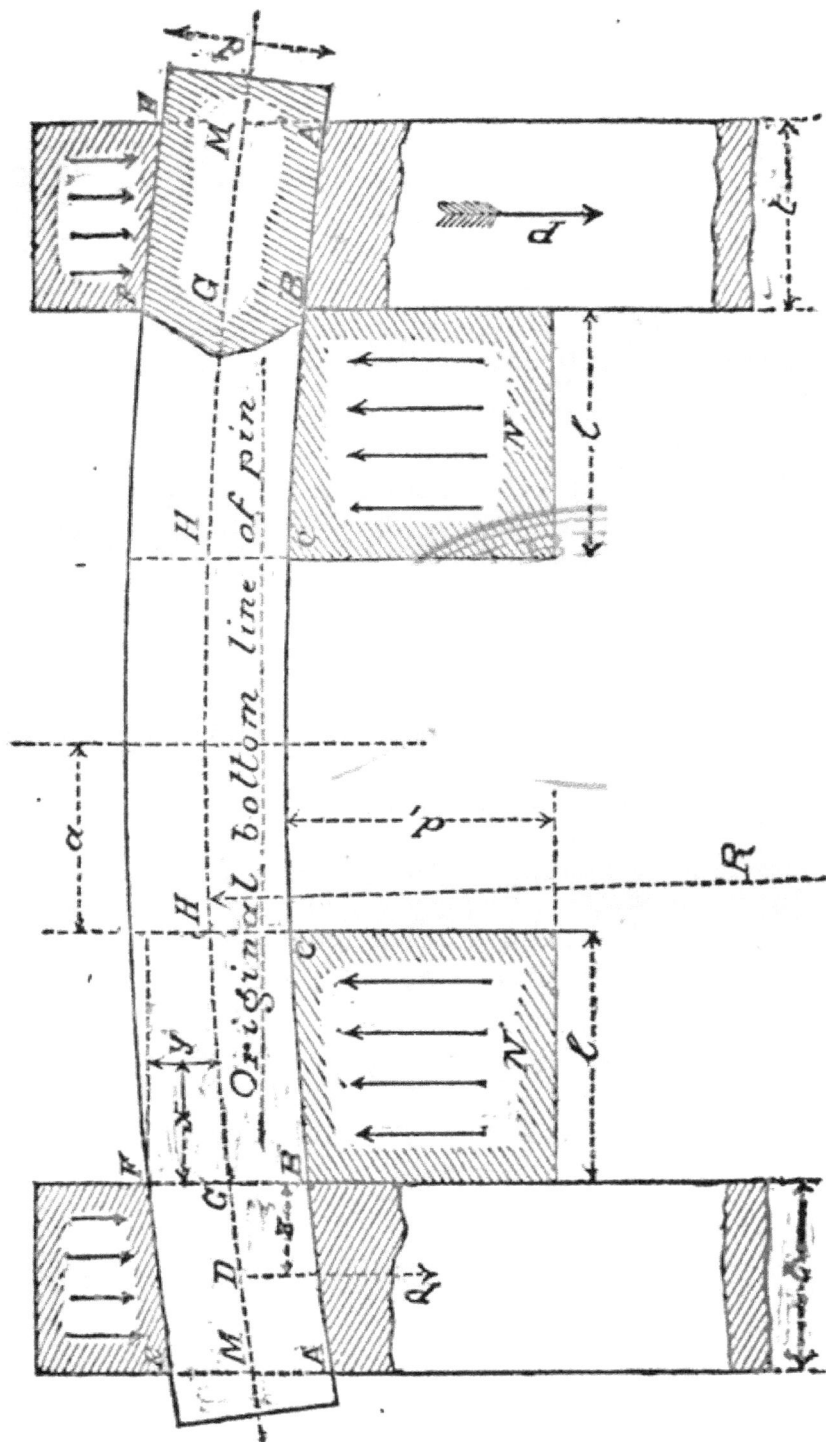

curve of depression. The curve itself consists of three parts ; the central part H H is circular, because the moment of flexure of H and H is of a constant value ; the part G H, which differs from H H as well as from M G, will be examined first. Take the origin of the co-ordinates in the surface B F, in the original centre line of the pin, which at G has received a depression Y o—and for any subsequent abscisses, has a deflection y.

The deflections y, are decreasing towards H, and consequently the first differential will have a negative value.

The cross section (Fig. 2) of the pin, represents how the deflection y is composed of two parts: one due to the compression of the pin itself, the other to the compression of the bearing, and fitting to each other. These partial depressions are $\frac{d}{2} \cdot \frac{A x}{E}$ and $d \cdot \frac{A x}{E}$, where A_x is the pressure per square inch, and E and E are the moduli of tensile elasticity of the material of which respectively the pin and bearing are composed.

The total depression consequently will be :

$$M M_1 = O p + p X = O X = \left(\frac{d}{2 E} + \frac{d_1}{E_1} \right) A x$$
$$= \frac{A x}{E} \left(\frac{d}{2} + d_1 \cdot \frac{E}{E_1} \right)$$

FIG. 2.

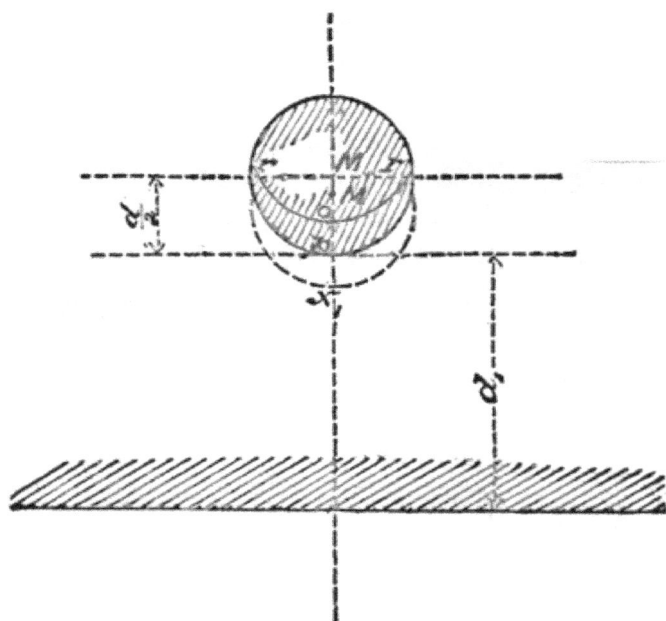

When the bearing is of cast-iron $\dfrac{E}{E_1}$ can be assumed to be about $= \dfrac{3}{2}$, so that

$$y = \frac{A\,x}{E}\left(\frac{d}{2} + d_1 \cdot \frac{3}{2}\right)$$

and if

$$\left(\frac{d}{2} + d_1 \cdot \frac{3}{2}\right) = \frac{d}{2} + d_1 \cdot \frac{E}{E_4}$$

is represented by L

There will be the equation

$$y = \frac{A\,x}{E} \cdot L \quad . \quad . \quad \text{Equation I.}$$

This equation also holds good for the pin-end M G, where l has only to be replaced by a proper part of the length of the bar.

The pressure A_z multiplied by the width of the bearing, represents the increment of the shearing force, taken per unit of the abscissa ; it therefore depends simply on the value y, and varies with the abscissa ; the law of the curve will be represented by an unknown function of y and X.

The fundamental law of any flexure of a beam is expressed by the equation

$$E\,I\,.\,\frac{d^2\,y}{d\,x^2} = M_x \quad . \quad . \quad . \quad \text{II.}$$

which expressed in words is—the product of the modulus of elasticity E, the moment of inertia I, of the section of the beam (here a pin), and the second differential coefficient of y is equal to the moment of exterior forces M_x taken at the point of abscisses considered.

It is known from the elementary theory of elasticity that the first differential coefficient of M_x is the shearing force for the point whose abscissa is x, and that its differential coefficient—the second differential coefficient of M_x is the increment of the shearing strain.

This increment can be derived from equation I., which gives $A_z\,.\,d = y\,.\,\dfrac{E\,d}{L}$, whence by taking the second differential of equation II., there is found

$$E\,I\,\frac{d^4\,y}{d_z{}^4} = \frac{d^2\,M_x}{d^z{}^2} = y\,.\,\frac{E\,d}{L}$$

or

$$\frac{d^4\,y}{d_x{}^4} = y \cdot \left(\frac{d}{I\,L}\right) \qquad . \qquad . \qquad \text{III.}$$

which integrated will give the equation of the curve. This is a lineal differential equation of the fourth order, and when

$$\frac{d}{I\,L} \text{ is called } p^4, \text{ where } p = \sqrt[4]{\frac{d}{I\,L}}$$

it will lead to the general formula : there being $e = 2.7182818$.

$$y = A_1\,e^{px} + A_{11}\,e^{-px} + B_1 \cos p\,x + B_{11} \sin p\,x. \quad \text{IV.}$$

for which the 4 constants must be determined to suit the conditions of the piece G H of the pin.

These are :

(a.) For $x = 0$ or at point a the moment of flexure is a known quantity $= \mathrm{P}\,z$ so that

$$\mathrm{E\,I}\,\frac{d^2\,y}{d\,x^2} = p^2\,\mathrm{E\,I}\,[A_1\,e^{px} + A_{11}\,e^{-px} - B_1 \cos p\,x$$
$$- B_{11} \sin p\,x]_{x=0} = \mathrm{P}\cdot z$$

or by inserting $x = 0$ in the formula

$$A + A_{11} - B_1 = \left(\frac{\mathrm{P}\,z}{p^2\cdot\mathrm{E\,I}}\right) = \mathrm{C}$$

(b.) For $x = 0$ the shearing force is equal to P, so that

$$\left[\mathrm{E\,I}\frac{d^3\,y}{d\,x^3}\right] \text{ for } x = 0 = p^3\mathrm{E\,I}\left[A_1 - A_{11} - B_{11}\right] = \mathrm{P}$$

$$\text{and } A_1 - A_{11} - B_{11} = \left(\frac{\mathrm{P}}{p^3\mathrm{E\,I}}\right) = \mathrm{C}_1.$$

(c.) At the end of the bearing the whole pressure P has been absorbed by the bearing casting, so that

$$\left[EI\frac{d^3y}{dx^3}\right] \text{ for } x = l = p^3 E\,I\left[A_1\,e^{pl} - A_1\,e^{-pl} + \right.$$

$$\left. B_1\,\sin pl - B_{11}\cos pl\right] = 0, \text{ and}$$

$$A_1 \cdot^{pl} - A_{11}e^{-pl} + B_1\,\sin pl - B_{11}\cos pl = 0.$$

(d.) The angle under which the tangent of the curve at H will cut the line of abscisses can be calculated thus:

The piece H H is part of a circle, whose highest point is midway between H and H_1, and if R represents the radius, then will

$$EI\frac{1}{R} = p^2\,EI(A_1\,e^{pl} + A_{11}\,e^{-pl} - B_1\cos pl -$$

$$B_{11}\sin pl)$$

and the tangent of the angle in question, since R is many times greater than $\frac{H\,H}{2} = a$, will be found $= \frac{a}{R}$, this being the absolute value of the tangent, whilst the first differential coefficient $\frac{dy}{dx}$ is negative (as has been already stated); then will

$$a\,p^2\,(A_1\,e^{pl} + A_{11}\,e^{-pl} - B_1\cos pl - B_{11}\sin pl)$$

$$= p\,(A_1\,e^{pl} - A_{11}e^{-pl} - B_1\sin pl + B_{11}\cos pl)$$

and

$$A_1\,e^{pl}(ap - 1) + A_{11}\,e^{-pl}(ap+1) - B_1\,(ap\cos pl +$$

$$\sin pl) - B_{11}\,(ap\sin pl - \cos pl) = 0.$$

The four equations a, b, c, and d are sufficient to calculate A_1, A_{11} B_1 and B_{11}, so that equation IV. is fully developed, and, according to equation

L. the pressure in any point of the bearing B C, can be found.

This somewhat intricate investigation was necessary to get an idea how the pressure P will be distributed. The subsequent example will illustrate the theory of this—

Let P $= 3'' \times 1'' \times 10000$ lbs. $= 30000$ lbs.

The lever z is very nearly $\frac{1}{4}$ of the thickness of the bar $= \frac{1}{2}''$. The diameter of pin $= 3''$, so that I$=4$. L can be assumed to be $= 4''$, $l = 1''$, E $= 30000000$ lbs. and $a = 5''$.

The value $p = \sqrt{\dfrac{3''}{4\times4}} = 0.658$ and $p\,l = 0.658$.

$C = \dfrac{z \cdot P}{p^2\,E\,I} = \dfrac{30000}{2\times0.658^2\times30000000\times4} = 0.0002887$

$C_1 = \dfrac{P}{p^3\,E\,I} = \dfrac{30000}{0.658^3\times30000000\times4} = 0.000877.$

$e^{pl} = 2.71828^{0.658} = 1.931.$

$e^{-pl} = 2.71828^{-0.658} = 0.5179.$

$\sin 0.658 = 0.6115.$

$\cos 0.658 = 0.7912.$

And these values put in the equations a, b, c, d, lead to

$$B_1 = -0.000902$$
$$B_{11} = -0.000165$$
$$A_1 = -0.0000493$$
$$A_{11} = -0.0006627.$$

For $x = 0$ there are

$y\,o = A_1 + A_{11} + B_1 = C + 2\,B_1 = -0.001615''$ and

$$\mathbf{A} \, o = y \, o \, \frac{E}{L} = \frac{30000000}{4''} \times 0.001615'' = 12115 \text{ lbs.}$$

For $x = l = 1''$ there are

$$y \, l = -0.001252'' \text{ and } \mathbf{A}_{l=1''} = 9391 \text{ lbs.}$$

If the pressure in the pin hole were all the constructor has to provide for, the dimensions of the pin and eye-bar might be determined in several ways. In reality, these dimensions are almost fixed, for, as will be shown, to reduce the flexure of the pin, the bearing surface must be short. The rule may be adopted to make the bearing surface B C as long as the eye-bar is thick. In this case, the pressure at B will be 12,115 lbs. whilst at C it is only 9,391 lbs. per square inch for a 3'' pin acted upon by a $3'' \times 1''$ eye-bar strained to 10,000 lbs. per square inch.

The maximum pressure is 21 per cent. greater than it would have been if uniformly distributed. The depression at B is 0.0016'' and at C 0.0012'', showing the influence of a curvature which hardly can be measured even by very fine tools, and generally would escape notice. The rise in the centre of the pin between H and H will be

less than 5 $\times$ (0.0016—0.0012) = 0.002″
—hence it is *unnecessary to provide an
upper bearing* in the centre of the casting
if the pin is nearly of the proper proportion; the play in the pin hole usually ex-
ceeds $\frac{1}{64}$ of an inch, which is more than six
times greater than the rise of pin between
H and H.

The maximum pressure for the standard
bearing length equal to one thickness of the
eye-bar was by the foregoing calculation
12,115 lbs.; for a badly fitting pin it would
be much larger, since then it is not uniformly distributed over the diameter of the
pin, but concentrated at one point. But
for a well fitting pin of large dameter the
pressure of 12,000 lbs. per square inch is
not too large; and for simplicity, it is well
to assume that this pressure is uniformly
distributed over the diameter of the pin,
until at least the effect of "play" in the
hole has been directly determined by a
large number of experiments on impact.

The later experiments prove conclusively,
that wrought-iron after millions of impacts
may break on the side where the strain is

tensile, but never on the side where the strain is compressive. Experiments recently made in this country as to the crushing strength of wrought-iron support this observation, the ultimate crushing strength having reached 60,000 lbs. per square inch. This quite disproves conclusions from older experiments carried up to ultimate strength, which led to the belief that iron under compression is weaker than under tension; which may perhaps be true for very soft metal. But such iron would not show the same behavior when used in a bridge. A properly proportioned bridge, having no section strained to more than 10,000 lbs. per square inch, will never break from softness of metal in compression, although it may after the passage of millions of trains by the ultimate failure or wearing out of its tension members. This view was always held by the best engineers on the Continent of Europe, and General Morin repeatedly expressed the opinion laid down here.

Fortunately skeleton bridges such as are built by reliable and experienced American engineers, can be calculated ; and all good ones are calculated so that no detail is strained nearly to the limit of durability.

To prove how specially erroneous are conclusions derived from Hodgkinson's experiments which refer to alleged differences in the compressive strength of wrought and cast iron, results obtained in France, and illustrated by General Morin, may be quoted.

Two cast-iron beams were made of the same metal and with the same height, length, and area. One of these beams was constructed to suit Hodgkinson's experiments with a heavy tension and a light compression flange, the two being in the ration of 4-7 to 1 ; the other had two equal flanges, the web was in both beams of the same thickness. The Hodgkinson beams deflected $2\frac{2}{3}$ times more than the plainer one ; this ratio according to the theory being exactly as that of the moments of inertia of the cross sections.

The Hodgkinson beam had to stand pressures more than 5 times greater than those of the other beam. The same results were arrived at by testing rolled wrought-iron beams with unequal flanges, which, when reversed, gave precisely the same deflections. Thus, these two experiments proved that the rules which Mr. Hodgkinson drew from tests up to ultimate strength were useless, to say the least, for parts strained below the elastic limit of the material. Again, to show how unreliable are rules derived from experiments carried up to ultimate strength, those of Mr. Fairbairn on a riveted girder are referred to. They were such as to expose the material to strains nearly in the same manner as for a railroad bridge. The girder broke under a strain of not more than 18,347 lbs. per sq. in. after only 5,175 impacts, and, of course, on the tensile part.

It must be supposed that this girder was of good workmanship, and at least of the average quality of English iron; and we know that a good wrought-iron bar *does not break* under strains of 30,000 lbs. *after* 130,000,000 of impacts.

The girder was repaired; when tested under a strain of 13,000 lbs. per sq. in., it did not break after 2,720,000 impacts. In accordance with new experiments, probably this girder would have broken after a sufficient number of impacts, still, evidently from the first experiments, it actually had *less than* 60 *per cent.* of what was *thought to be the available area.*

Since in such girders generally about 20 per cent. of material is wasted in rivet holes, it may be said, though properly designed and constructed according to the *rules* derived from experiments on *ultimate strength*, that when tested in the same manner as in practice, they have *lost more than half* of the value of their metal. The failure of the Crumlin Viaduct superstructure is another and more direct illustration in proof that experiments to the ultimate strength cannot be relied on in deducing rules for the proportions of pins in bridges.

This failure was due to pins proportioned by a rule derived from experiments on the ultimate shearing strength of rivets. Such rules, applicable to boilers or ships where the ultimate strength must be taken into account, cannot be safely used to determine the proper diameter of a pin. One of their defects is that no allowance is made for pressure in the hole, which frequently is $2\frac{1}{2}$ times the strain calculated to be uniformly distributed over the cross section.

Rivets, on account of friction caused by their heads, transfer a portion of this pressure to the outer surface at the plates they join together, and therefore do not give strikingly bad results in practice.

The same rules cannot apply to pins where no such pressure on the surface takes place.

The deduction of rules for pins from the condition of rivets is not the only empiricism in regard to the former. In Germany, on occasion of the erection of suspension bridges, Engineer Malberg made trials of links which gave results agreeing with those obtained later in England. Though the German proportions were published, it seems they did not receive attention abroad, or else the English experiments would probably not have been made. As far as value is concerned, neither set of trials should be relied on.

Specifications for our bridges require

consideration only of a *maximum direct strain* under the heaviest load which the structure may bear. This condition, with reference to pins, can readily be fulfilled by examining analytically the nature of their strains. A pin is nothing but a beam, and since a great variety of experiments, made on beams, prove that within the limits of elasticity the theory adopted is not less correct than that of the law of gravitation to the movements of the planets, what remains to be done is: only to find theoretically the maximum strains at different points of the pin. Recourse, however, was had to empirical researches, and tests were made which could not show the nature of the strains under loads such as occur in practice. Thus the first experimental English rule made no allowance for pressure and flexure, but referred solely for the shearing strain which was supposed to be uniformly distributed over the cross section of the pin. It is plain that at the elastic limit the science of strains ends, since this depends on the principle, "ut tensio sic vis."

Nevertheless this is still frequently overlooked, and especially so in reference to the shearing strain of pins.

The rule, moreover, has been frequently misapplied as to pins of suspension bridge chains, which have been considered as subjected to double shearing, although the outer bars always cause but single shearing, and also not unfrequently to bottom chord pins, where likewise the outside bars cause but single shearing.

This rule regarding exclusively the shearing strength of a pin was used until the failure of the Crumlin Viaduct, and experience gained with suspension bridges (such as built at Montrose in Scotland, where the pins in a few years cut their way almost through the eyes) caused engineers to make other trials referring to the strength of eyes and the bearing surface of pins. These experiments were with wide and thin bars, as used in suspension bridges, but not in truss bridges, of good design. In this case the eye of the bar, placed between two links or the jaws of the machine, acts on the pin by double shear; the action is the same with a bar as wide as the one tested but one half as thick, placed outside of the bottom chord of a truss bridge.

The new rule deduced, fixing the diameter of the pin from $\frac{2}{3}$ to $\frac{3}{4}$ of the width of the eye bar, contains no provision for the *thickness* of the bar, and applies to the case where the bar is 20 times as wide as thick and the pin is subjected to single shearing. Whether this rule applies to a square bar is more than doubtful for two reasons. First: Under a test up to ultimate strength the pin will flatten, the bearing surface of the eye will be increased and receive a remarkable permanent set, and the now tightly fitting pin will exert a great radial pressure on the pin hole, which causes friction that may be nearly as great as itself, since under high pressure friction increases greatly.

A pin in a bridge is used quite differently; the bearing surface is much less, the pin hole will flatten but little, and the pin would wear out the hole in a comparatively short time if made to conform solely to experiments on the ultimate strength. It seems to follow from such experiments that the pressure on the bearing surface could be considered as uniformly distributed over the semicircular surface instead of through the diameter of the pin. This cannot be correct for a pin which in practice has a play of $\frac{1}{64}$th to $\frac{1}{32}$d of an inch, and which ought not to be pressed more than about 12,000 lbs. per sq. in. Second : The latest English rule does not take account of the thickness of the bar, and of the moment of flexure to which the pin is exposed. This has been already alluded to, and is really the leading point in determining the size of a pin ; the dimensions which satisfy this condition will also satisfy the two others.

Having thus explained why experimental reseaches have not as yet established practicable rules for pins, the examination of shearing strains with reference to pins, a subject which has not been sufficiently discussed in many text-books, will be considered.

SHEARING STRAIN IN PINS.

The theory of flexure teaches that the shearing strain is not uniformly distributed over the cross section ; consequently the

maximum shearing strain must exceed the
strains which could exist if uniformly distributed. To show the amount of their
difference: Figure 3 represents a part of a
bent beam; C E and D E₁ must be two
imaginary sections across it, and V V, a surface parallel to the neutral surface G B.

FIG. 3.

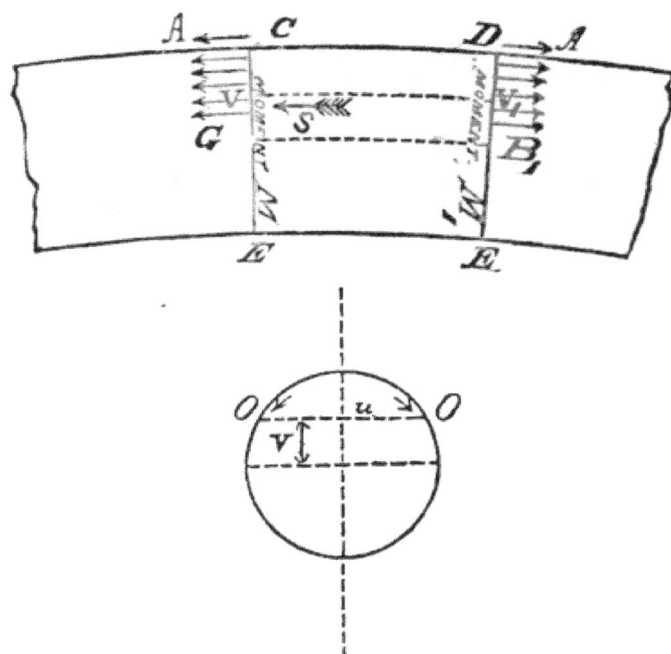

The question is what forces keep the
body G D V V₁ in equilibrium. The moment of flexure at A generally differs from

the moment at B, which, in this examination, is supposed to be the greater. The consequence is that the maximum strain per square inch at D is greater than at C. These strains per square inch at D and C may be represented by the letters A_1 and A.

The strain per square inch in any point V is $A \times \dfrac{\bar{v}\,G}{C\,G}$, since the strains decrease in the same ratio as the point approaches the neutral line. The sum of all the strains of the surface C V is partly counteracted by those of the surface D V, the last named sum being the greater. It consequently needs a shearing force S, acting in the direction of A, to resist the forces in the direction of A_1.

Represent by V any distance G V and by u the width $\overline{o\,o}$, then that for the distance V=VG and the total shearing strain can be represented by

$$S = \frac{A_1 - A_1}{(C\,G)} \int_{V\,=\,G\,V}^{V\,=\,G\,C} (v\,u\,d\,v)$$

If the sections C G and D B approach each other closely, the shearing strain must be considered as being uniformly distributed over the surface VV′ and the strain per square inch will be;

$$\sigma = \frac{S}{\overline{vv'oo}} = \frac{A_1 - A}{\overline{vv}} \cdot \frac{1}{c\,G.u} \cdot \int_{v=Gv}^{V=CG=a} (u\,v\,d\,v)$$

The values A and A_1 can be derived from their respective moments M and M_1 so that when I represents the moment of inertia of the cross-section of the beam there will be $A_1 = \frac{a}{I}$. M_1 and $A = \frac{a}{I}$. M_1, and the shearing strain per square inch will be:

$$\sigma = \frac{I}{I \cdot u} \cdot \frac{M_1 - M}{\overline{v\,v^1}} \cdot \int_{v}^{a} (u\,v\,d\,v)$$

and $\frac{M_1 - M}{\overline{v\,v^1}} = \frac{d\,m}{d\,x} = V =$ the total shearing force for the section C E or D F, so that the shearing strain per square unit will be

$$\sigma = \frac{V}{I\,u} \cdot \int_{v}^{a} (u\,v\,d\,u)$$

For $v = 0$, σ will be a maximum, which is the case for the neutral line itself, whilst the integral decreases to nothing at the point D. For a pin, the section is a circle, and $u^2 + 4\,v^2 = d^2$, d being the diameter of the pin, and the maximum shearing strain will be found:

$$\sigma = \frac{4}{3} \frac{V}{(r^2\,\pi)}$$

so that σ is $1\frac{1}{3}$ times larger than if the total shearing force V were uniformly distributed over the cross-section $r^2 \pi$ of the pin.

The higher and more accurate examination proves that the shearing strain also is not uniformly distributed over the lines *oo*, that the absolute maximum shearing strain is in the centre of the pin, and is exactly $1\frac{2}{3}$ times larger than it would be if the shearing force were uniformly distributed over the whole section.

There is but a step from the longitudinal to the vertical shearing strain. That both are equal in every point of the beam, (on the pin) can readily be seen by considering the infinitely small hexahedron of Figure 4.

This body could not remain in equilibrium if the horizontal shearing force, σ, were not prevented from turning the figure $\overline{A B} \overline{C} D$ around point B, by the vertical shearing force σ, which works with the lever B A, to turn the figure A B C D the other way. A B being $=$ B C $=$ the units, the vertical and horizontal shearing force must be equal. This law is a general one suitable for any body.

Now it has been proved that the shearing maxima strains which act in the centre of a pin, both horizontally and vertically, are $1\frac{3}{8}$ times larger than if it were possible to distribute the total shearing force V uniformly over the cross-section. The value V reaches its maximum just at the facing of the casting, and is equal to the total tension of the eye-bar.

FIG. 4.

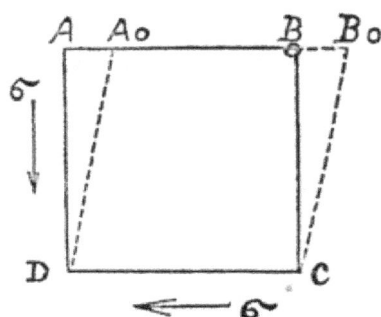

How large can the shearing strain σ be, without exceeding the ordinary requirement for iron bridges, that no tensile strain shall be greater than 10,000 lbs. per square inch? In answering this, attention is called to Figure 4.

Shearing strain is but a force which tends to slide an infinitely thin slice of a body along its section, as for instance, the surface A B parallel to C D, into the new position $A_0 B_0$. The absolute value of the deplacement $A A_0$ depends on different causes :

Firstly on the shearing strain σ itself.

Secondly on the distance A D between $\overline{A B}$ and $\overline{C D}$, for a surface parallel to C D, midway between A and D would only slide half as far as A B. This law is true, since within the limits of elasticity all displacements increase in direct and single ratio with the lengths.

Thirdly on the nature of the material, so that

$$\overline{A A}o = \sigma \times A D \times \left(\frac{1}{G}\right) \text{ where } \left(\frac{1}{G}\right) \text{ is a co-}$$

efficient dependent on the nature of the material,

or it is $\dfrac{\overline{A A}o}{A D} = \dfrac{\sigma}{G}$ tang. angle (A D A o).

Since σ is a finite value like G, the angle $A D A_0$ is also of a definite value. This angle is the test of shearing strain, so that wherever an angle has changed from right, a pair of shearing strains must have caused it.

The value G must represent a *weight* in order to make $\frac{\sigma}{G}$ an abstract fraction. G is called the modulus of shearing elasticity, and represents the weight which, if the limits of elasticity would reach thus far, were sufficient to slide a surface so far that the angle A D A$_0$ would become 45 deg.

A shearing strain can always be resolved into tensions and compressions acting in all possible directions on a point in the interior of a strained body.

FIG. 5.

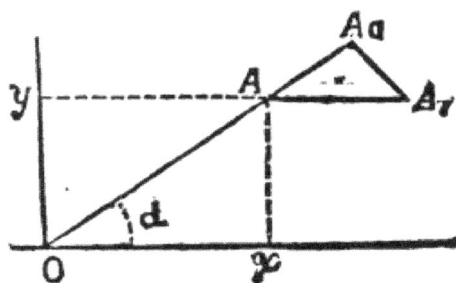

In Figure 5, O represents a point of a body, A A$_1$ the shearing displacement of the infinitely near point A. O x parallel A A$_1$ is made first axis of the system of co-ordinates.

The original line O A, by the shearing

strain has been changed into O A$_1$, which is equal to OA plus AA$_o$, this being the projection of A.A$_1$ on OA.

The shearing strain per square inch being σ, there will be

$$A A_1 = y \cdot \frac{\sigma}{G} \text{ and } A A o = y \cdot \frac{\sigma}{G} \cdot \cos a$$

and

$$\frac{A A o}{O A} = \frac{\text{tension}}{\text{modulus}} = \frac{y}{O A} \frac{\sigma}{G} \cos \alpha$$

or

$$T \sin \alpha \cdot \cos \alpha \cdot \frac{E}{G} \sigma$$

This value will be a maximum when α =45 deg., so that Tension maximum

$$= \tfrac{1}{2} \cdot \frac{E}{G \, c} \cdot \sigma$$

For a perfectly homogeneous body by experiment and calculation, G is found to $= \tfrac{2}{5}$ E, so that the maximum tension which accompanies any shearing strain in such a body (good iron or steel, but not wood) will $= \tfrac{1}{2} \cdot \frac{E}{\frac{2}{5} E} - \sigma = \frac{5}{4}, \sigma$. or if the maximum tension is limited to 10,000 lbs. per sq. in., no shearing strain must be greater than 8,000 lbs. per sq. in.

It has been shown that in the centre of any pin the shearing strain is $1\frac{3}{8}$ times greater than if the shearing force were uniformly distributed over the cross section, hence the pin must be proportioned to withstand a uniformly distributed shearing force of $1\frac{3}{8}$ times the actual one; in other words, the shearing strain must only be $\dfrac{8000}{1\frac{3}{8}}$ lbs.= 5,810 lbs. per sq. in.*

If this condition is observed the tension in the centre of the pin acting at 45 deg. to its axis will be not more than 10,000 lbs. per sq. in.

If the section of the tie bar is "b" by "t" the total shearing force will be $10,000 \times b \times t$ and the section of the pin will be $\dfrac{10000\,b\,t}{5820}$ =1.72 $b\ t$., or the section of the pin must be nearly equal $1\frac{3}{4}$ *the section of the bar.*

This condition determines that in the neutral axis of the pin the shearing strains, tensions, and compressions shall not exceed

---

* The Baltimore Bridge Company make the uniformly distributed shearing strain= 6,000 lbs. per sq. in.

the maximum tension usually prescribed, of 10,000 lbs. per sq. in.

The rule would apply if another amount, as 12,000 or 15,000, were prescribed. The limit of shearing strain should in such a case be raised correspondingly, by still making the pin section $1\frac{3}{4}$ times the section of the bar.

It will, however, be seen that consideration of the shearing strain alone is not sufficient to properly proportion a pin.

The results obtained thus far depend on the modulus, G being $\frac{2}{5}$ of E for a homogeneous body as good iron or steel. This assertion must now be proved. By purely mathematical investigation, Navier, then Cauchy, Dienger, and others found that any pressure on a perfectly homogeneous body is accompanied by an expansion or lateral tension equal to $\frac{1}{4}$ of the pressure per sq. in., and the tension accompanied by lateral compression equal to $\frac{1}{4}$ the value of the tension. Rude experiments with india-rubber prove the existence of lateral compression or tension, and those made by Wertheim and Regnauld

confirm the theory sufficiently, the coefficient differing somewhat with the degree of
homogeneity of the bodies under test. For
iron, Wertheim found $\frac{1}{4}$, and sometimes a
little more, but not so much as to cause a
change in the modulus to exceed $1\frac{1}{2}$ per
cent. Therefore, without entering into the
analytical investigations of Navier, etc., it
may be assumed that for well-rolled wrought
iron the lateral contraction or expansion is
$\frac{1}{4}$ the longitudinal tension or pressure.

FIG. 6.

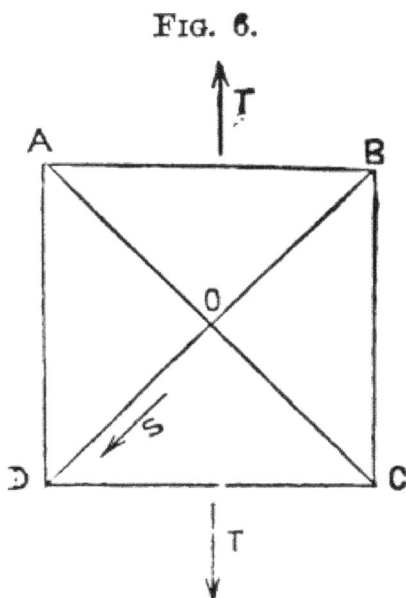

Fig. 6 represents one surface of a hexahedron, which of infinitely small sides is assumed to be acted upon by two tensional forces T, equally distributed over the surfaces indicated in the figure by the lines $\overline{AB}$ and $\overline{CD}$.

The sides $\overline{AD}$ and $\overline{BC}$ are extended and the sides $\overline{AB}$ and $\overline{CD}$ compressed. The extension will be $\dfrac{T}{E}$ BC. the compression $\dfrac{T}{4E}$ AB, or since AB=BC, the side itself can be assumed as unit, and the extension will be $\dfrac{T}{E}$ whilst the lateral contraction is $\frac{1}{4}$ of this value.

The angle $\overline{BOC}$ was originally 90 deg., but now it is more and the increase may be represented by the letter ε. If B O C increase, the angles O C B and O B C decrease each one-half the amount of E.

There is consequently after the extension

$$\tan \left(45° - \frac{\varepsilon}{2}\right) = \frac{1 - \frac{1}{4}\dfrac{T}{E}}{1 + \dfrac{T}{E}} =$$

$$\frac{\text{A B after the contraction}}{\text{B C after the extension}}.$$

or since ε is a very small angle

$$\frac{\text{tang } 45° - \text{tang } \frac{\varepsilon}{2}}{1 + \text{tang } 45° \text{tang} \frac{\varepsilon}{2}} = \frac{1 - \frac{\varepsilon}{2}}{1 + \frac{\varepsilon}{2}} = \frac{1 - \frac{T}{4E}}{1 + \frac{T}{E}}$$

From this again (because ε is very small)

$$\left(1 - \frac{\varepsilon}{2}\right)\left(1 - \frac{\varepsilon}{2}\right) = \left(1 - \frac{T}{4E}\right)\left(1 - \frac{T}{E}\right)$$

or

$$1 - \varepsilon = 1 - \frac{5}{4}\frac{T}{E} \text{ and } \varepsilon = \frac{5}{4}\frac{T}{E}.$$

The angle A O B having changed its value, there must be a *shearing force along both diagonals.*

This shearing force being S_1 ε will be $= \frac{S}{G}$ and $\frac{S}{G}$

$$= \frac{5}{4}\frac{T}{E}.$$

The value S can also be found by a second consideration. If the prism A B D is in equilibrium, the force T, will be counteracted by a shearing force S, and a tensional force perpendicular to diagonal B D.

The projection of the total force acting on the surface, which is represented by the line A B, on the diagonal B D, must equal the total shearing force acting on the surface of the prism represented by line B D, or

$$1 \times 1 \times T \times \cos 45° = \overline{B\,D} \times 1 \times S, \text{ or}$$

$$T \times \cos 45° = \frac{1 \times S}{\cos 45°} \text{ or } T = 2\,S.$$

The two equations for S combined give the value of G represented by E, so that there is:

$$\frac{S}{G} = \frac{5}{4} \cdot \frac{2\,S}{E} \text{ or } G = \frac{2}{5}\,E -$$

This is the equation upon which the rule was based that the section of a pin must be at least $1\frac{3}{4}$ times the section of the bar, to keep the tension caused by the shearing strain below the limit generally prescribed. To many engineers the above deductions may be new, and it is therefore desirable to dwell for a few moments on the conclusions which may be derived therefrom. It has been mentioned that Wertheim and Regnauld made several series of experiments, by which they established that the factor of lateral contraction does most sufficiently correspond with the moduli E and G, as found by tensional and torsional experiments.

The theory of strains has received very valuable proofs by Chief Engineer Woehler, in Germany. He made experiments on impact, on tension, compression, torsion, etc., during 12 years. He first established the law that any material may be broken by repeating sufficiently often extensions, which, however, cause strains below the breaking point, and he then determined certain limits of strain within which the material did not break. Further experiments on torsion and other more direct chearing strains confirmed the same law, but established that this limit must necessarily be lower then for direct strains.

We may call these limits the "limits of durability."

Cast-steel, cut from railway axles furnished by Krupp, has not broken after 40,000,-000 impacts, straining the material transversely to 53,000 lbs. Nor could this steel be broken by any number of impacts causing shearing strains of 42,000 lbs. per square inch. All strains higher than these produced rupture after a sufficient number of impacts; and there seemed to exist a certain relation between the number of impacts and the value of the strain. We may therefore call 53,000 lbs. the limit of durability for tensile, and 42,000 lbs. the limit for shearing strain of the class of steel experimented on.

According to theory, the ratio between both limits ought to be as 1 : 0.800, whilst in reality it was as 1 : 0.793.

The tensile modulus of this steel was experimented on, and fixed at 28,725,000 lbs., while the shearing modulus, found by very careful torsional experiments, was 11,237,-000 lbs. Their ratio, according to theory, ought to be as 5 : 2.—when the result of the experiment proved it to be as 5 : 1.95.

The quoted results refer only to a very small part of the experiments, which were extended to copper, and wrought and cast-iron of different makes ; all gave results in good harmony with theory, and as they were made to test the material in the same manner as in practice, they prove that, within the limits for which the theory holds good, we can well rely on it with this proviso—that for riveted, forged or machine-worked parts, experiments should be made of each class of material, by which alone a correct idea as to their strength and endurance can be formed. The older experiments, made on strains beyond the limits of durability of the material, or not with acting in the same manner as in practice, or carried to the ultimate point of failure, could not possibly lead to any law or reliable formula, because the phenomena beyond the elastic limit cannot be followed up by even the highest analysis, and possibly do not conform to any law.

Frequently, as a consequence of the older views, the effects of shearing, torsional, compressive, and tensile strains and strength, are spoken of as so many different phenomena without relation to each other, while in reality in every point of any strained body there exist shearing, tensile, and compressive strains at the same time.

Referring to the object of this paper, it may be added that experiments on the ultimate shearing strength show that the results greatly depend on the way they are obtained, in some cases the shearing strength being found as great as the tensile strength, and in other cases only two-thirds of it.

If the greatest tension in a wrought-iron

bridge is assumed at 10,000 lbs. per square inch, the greatest shearing strain ought not to exceed 8,000 lbs. This is far within the limit of durability of good iron under tension, which after 132,000,000 of impacts was established at 33,000 lbs. per square inch.

Having now determined the fundamental laws of elasticity which enter into the problem of the shearing strains in a pin, and also having given the data necessary in estimating the maximum pressure that can be allowed in the pin-hole, a more important consideration remains, the nature of the strain caused by the moment of flexure of the pin is still to be examined.

Attention is called to Figure 1, representing the strains upon a pin either in a top chord of a truss bridge, at the exterior links of a bottom chord, or of chain suspension bridges. The question is where to find the maximum moment of flexure?

It is a very common mistake to assume that the maximum moment is in section F B, when, instead, it occurs in section H C. For by reversing Fig. 1 it will be seen that the problem is the same as if a beam, G G, were loaded on both sides from G to H, leaving a space H H unloaded, and that the two eye-bars form the supports of such a beam.

Consequently the moment must have a constant value from H to H, and the maximum moment must exist in any section of the pin between H and H.

The pressure on B C for the present purpose can be considered as uniformly distributed, and the strain of the eye-bar assumed as being concentrated in D, which is in the middle of line M G. The reaction likewise is then concentrated in the middle of B C, which, according to previous considerations, is made to be equal to M G.

In fact the centre of the reactionary forces is a little closer to B than to C, but the difference is very small, as long as the bearing is not longer than the bar is thick.

This supposition leads to the smallest possible pin, and when the bearing surface is made longer, the pin must be stronger, while the pressure in the pin hole at the same time decreases, but is less uniformly distributed.

The moment of flexure of the pin in the section H C, is $= P \left(\dfrac{M\,G}{2} + G\,H \right)$ less P. $\dfrac{G\,H}{2}$, or since G H was made $= M\,G$ there will be the maximum moment $=$ P . M G $=$ the strain of the eye-bar multiplied by the thickness of the bar as lever.

The exact value of the moment can be found by determining the value E I $\dfrac{d\,y^2}{d\,x^2}$ of the curve B C, for $x = l = $ B C of which the equation has been developed.

This value for *any* length P of B C would give the exact moment and strain which a pin has to bear, its diameter being assumed previously

The value P $\dfrac{M\,G + G\,H}{2}$ is much larger than it would have been obtained had section F B been the one examined. Then the moment would have been only P . $\dfrac{M\,G}{2}$, which is less by P . $\dfrac{G\,H}{2}$ than the actual maximum moment.

Under the assumption that M G $=$ G H the error is 100 per cent. and when G H$>$

M G it is still greater.   If the size of the pin
were calculated under this erroneous supposition, the actual strain would be twice
or more *than twice as large* as was intended.
There is no shearing strain and no compression by reaction in the section C H and
the tension in the uppermost fibres could be
used for the determination of the size of pins.

The well known elements of flexure prescribe that
there must be, for $l = t : \dfrac{S}{4}\left(\dfrac{d}{2}\right)^3 .3.14 = P\,t$, or
since $P = b \times t \times 10000$ lbs. and when S is fixed at
10000 lbs. per sq. in. $\dfrac{d^3}{32}$. $3.14 = b \times t^2$ and $d^3 =$
$10.18 \cdot b\,t^2$.

If $b = 4t$ $d = t^3\sqrt{40.72} = 3.44\,t = 0.86$ width of bar.

$b = 3l$ $d = t^3\sqrt{30.54} = 3.18\,t = 1.06$ "   "

$b = 2\,t$ $d = t^3\sqrt{20.36} = 2.73\,t = 1.36$ "   "

$b = t$ $d = \sqrt[3]{10.18} = 2.16\,t = 2.16$ "   "

$b = t$ (round bar) $d = t \ \sqrt[3]{8} = 2.00\,t = 2.00$
thickness of bar.

These sizes of pins are so large as to ex-
clude entirely the consideration of shearing
strain and of pressure in the pin-hole, but
it may be justifiable to permit greater strains
from flexure than has been prescribed for
the maximum tension of the tie-bar.   In-

deed there are reasons for this consideration. 1st. It is impossible to equally strain all bottom chord bars to exactly the specified limit; some bars will receive tension exceeding 10,000 lbs. per square inch, as will be explained hereafter. Also when several bars are fixed to the same pin it cannot be expected that they severally will have equal moduli of elasticity, hence the prescribed maximum of 10,000 lbs. may be exceeded.

2d. The eye-bars have been shaped by a process of manufacture, which under all circumstances somewhat impairs the uniform quality of the iron.

3d. A pin has not been exposed to fire after having been rolled; if turned to size with good machinery, and by a skilful workman, its quality has not been altered; hence the iron of which it is made is more reliable than that of any other part of the structure.

Under these suppositions, it is thought safe to permit a maximum tension of 12,000 lbs. per square inch, which is not more than

the maximum pressure in the pin-hole as
determined in the first part of this paper.

The results of the calculation brought to
practical dimensions, are represented in the
following table :

Width of bar......b 4″ 3″ 2″ 1″ 1″.
Thickness of bar..t 1″ 1″ 1″ 1 square 1 round.
Diameter of pin...d 3¼″ 3″ 2½″ 2″ 1¾″
Ratio of $d : b$.....0.8, 1.0, 1¼ to 1⅜, 2, 1¾ to 1⅞

The maximum shearing strain of these
pins is 6,600 lbs. per square inch—the
minimum about 5,000 lbs.

According to these results the old rule
of $\dfrac{d}{b} = \tfrac{3}{4}$ is only admissible for bars with
thickness ⅕ of their width, or less than this.
The use of this rule for a square bar would
permit a maximum strain of more than
60,000 lbs. per square inch, if the conditions did not change by the action of the
nut when a permanent set takes place, and
a pin of 1½″ diameter for a 2″ × 1″ bar,
would permit a maximum tension of nearly
50,000 lbs.

These two examples are sufficient to show
the importance of properly proportioned

pins, and that the best American bridge constructors are wise in using dimensions varying but little from those found above.*

Indeed for a considerable time the more scientific engineers of this country have strengthened by degrees their bridge connections, and by combining practical experience with theoretical considerations the theoretical dimensions have been nearly reached. European engineers by using riveted work have not had an opportunity to obtain this result. This affords a reason why in many late books the subject of pins is not treated more scientifically.

The practical bridge builder, when designing a structure, sometimes finds it difficult to fulfil the conditions of theory and manufacture which more or less contradict each other. However, the heavier tie bars which carry the greater part of the dead load are not so much influenced by the live load, which is the really destructive element,

* Mr. Thos. C. Clarke, of Philadelphia, stated in a recent letter to the American Society of Civil Engineers, that in the bridges of the Phoenix Bridge Co., nearly the same diameters as given in the last table are used.

while the pins of the lighter bars, near the centre of the top chord, are generally stronger than necessary.

The diameter of top chord pins can be reduced by making them what is called double shearing, a bearing being placed on each side of each bar; the diameter can then be determined as if the bar had only $\frac{6}{10}$ at its thickness.

It has been seen how necessary it is to use the proper size of pin in the top chords, which is more especially true for pins bearing more than one bar. In this case the sum of the bending moments produced by the several bars must be introduced in the formula for the tensile strain by flexure.

Great care is also required in properly proportioning the pins of bottom chords of bridges. Generally here the pin takes hold of the post, directly to it are attached the ties, outside of which the bottom bars are arranged. Here the greatest moment of the pin is near the bottom bar next to the tie. It is an accumulation arising from the sometimes great number of bars. In order to form a more precise idea of the strains which may arise, it may be as well to examine a practical example.

Suppose the three equally strained bottom bars 3 in.$\times$1 in., 3 bars 3 in.$\times$1¼ in., and a tie 3 in.$\times$1 in. are on the same pin on one side of the post of the bridge. The maximum moment will be 5.1 in.$\times$10,000 lbs., and the maximum tensional strain by flexure 12,100 lbs., whilst the pressure in the pin hole will be found to be 8,570 lbs., resulting in a tension of 14,200 lbs.

Some bridge-builders use square bars, with pins not much larger than the thickness of the bar.

Suppose, in reference to the above example, that three 2 in. square bottom bars were counteracted by three 1¾ in. square bars and the tie. For a 3 in. pin the tensile strain by flexure would amount to 25,-000 lbs. Of course neither this nor the above strain of 14,200 lbs. will really exist in the pin. The fact will be that the pins bend a little toward the centre of the bridge, and the bars near to the centre line of the bottom chord will bear a greater share of the chord strain, relieving the outer bars.

The difference will be the greater, the smaller the pin; the greater the difference in size of bars, the thicker the bars, and the shorter will be the panels.

The difficulty of reducing the strains of bottom pins to the strain of the bars increases with the magnitude of the bridge, and may partly be met by increasing the number of flat and thin bars, making up for the difference in tension in two adjoining panels rather by the number than by the size of the bars, and especially by using the proper size of pin.

Another way to meet the difficulty consists in creating two centre lines in the bottom chord, by placing a proper number of chord bars between the ties, by using two posts, or by constructing a properly built post foot. For small bridges these costly arrangements can be dispensed with. Flat bars, in preference to round or square ones, are in all cases to be recommended. The Baltimore Bridge Company, for instance, always use such, and it seems that other constructors have recently adopted the same, to the exclusion of round bars.

Pins used in chain suspension bridges, the bars being all of equal section, have to resist only a small moment; their proportions follow the rule for top pins.

As a conclusion to which the foregoing examination perhaps has led, it may be mentioned that it is easier to calculate the general strains of skeleton structures than to design details, which, satisfying the practical demands of economical, speedy, and reliable manufacture, are also in harmony with the more subtle ones concerning their proportions.

VALUABLE
SCIENTIFIC BOOKS,

PUBLISHED BY

D. VAN NOSTRAND,

23 MURRAY STREET AND 27 WARREN STREET,

NEW YORK.

FRANCIS. Lowell Hydraulic Experiments, being a selection from Experiments on Hydraulic Motors, on the Flow of Water over Weirs, in Open Canals of Uniform Rectangular Section, and through submerged Orifices and diverging Tubes. Made at Lowell, Massachusetts. By James B. Francis, C. E. 2d edition, revised and enlarged, with many new experiments, and illustrated with twenty-three copperplate engravings. 1 vol. 4to, cloth......................$15 00

ROEBLING (J. A.) Long and Short Span Railway Bridges. By John A. Roebling, C. E. Illustrated with large copperplate engravings of plans and views. Imperial folio, cloth.............................. 25 00

CLARKE (T. C.) Description of the Iron Railway Bridge over the Mississippi River, at Quincy, Illinois. Thomas Curtis Clarke, Chief Engineer. Illustrated with 21 lithographed plans. 1 vol. 4to, cloth...... .... 7 50

TUNNER (P.) A Treatise on Roll-Turning for the Manufacture of Iron. By Peter Tunner. Translated and adapted by John B. Pearse, of the Penn-

I

sylvania Steel Works, with numerous engravings
wood cuts and folio atlas of plates.................. $10 00

ISHERWOOD (B. F.) Engineering Precedents for
Steam Machinery. Arranged in the most practical
and useful manner for Engineers. By B. F. Isherwood, Civil Engineer, U. S. Navy. With Illustrations. Two volumes in one. 8vo, cloth........... $2 50

BAUERMAN. Treatise on the Metallurgy of Iron,
containing outlines of the History of Iron Manufacture, methods of Assay, and analysis of Iron Ores,
processes of manufacture of Iron and Steel, etc., etc.
By H. Bauerman. First American edition. Revised
and enlarged, with an Appendix on the Martin Process for making Steel, from the report of Abram S.
Hewitt. Illustrated with numerous wood engravings.
12mo, cloth.. 2 00

CAMPIN on the Construction of Iron Roofs. By
Francis Campin. 8vo, with plates, cloth.......... 3 00

COLLINS. The Private Book of Useful Alloys and
Memoranda for Goldsmiths, Jewellers, &c. By
James E. Collins. 18mo, cloth.................... 75

CIPHER AND SECRET LETTER AND TELE-
GRAPHIC CODE, with Hogg's Improvements.
The most perfect secret code ever invented or dis-
covered. Impossible to read without the key. By
C. S. Larrabee. 18mo, cloth..................... 1 00

COLBURN. The Gas Works of London. By Zerah
Colburn, C. E. 1 vol. 12mo, boards.............. 60

CRAIG (B. F.) Weights and Measures. An account
of the Decimal System, with Tables of Conversion
for Commercial and Scientific Uses. By B. F. Craig,
M.D. 1 vol. square 32mo, limp cloth............. 50

NUGENT. Treatise on Optics; or, Light and Sight,
theoretically and practically treated; with the application to Fine Art and Industrial Pursuits. By E.
Nugent. With one hundred and three illustrations.
12mo, cloth.. 2 00

GLYNN (J.) Treatise on the Power of Water, as applied to drive Flour Mills, and to give motion to
Turbines and other Hydrostatic Engines. By Joseph

Glynn. Third edition, revised and enlarged, with numerous illustrations. 12mo, cloth.............. $1 00

HUMBER. A Handy Book for the Calculation of Strains in Girders and similar Structures, and their Strength, consisting of Formulæ and corresponding Diagrams, with numerous details for practical application. By William Humber. 12mo, fully illustrated, cloth.. 2 50

GRUNER. The Manufacture of Steel. By M. L. Gruner. Translated from the French, by Lenox Smith, with an appendix on the Bessamer process in the United States, by the translator. Illustrated by Lithographed drawings and wood cuts. 8vo, cloth.. 3 50

AUCHINCLOSS. Link and Valve Motions Simplified. Illustrated with 37 wood-cuts, and 21 lithographic plates, together with a Travel Scale, and numerous useful Tables. By W. S. Auchincloss. 8vo, cloth.. 3 00

VAN BUREN. Investigations of Formulas, for the strength of the Iron parts of Steam Machinery. By J. D. Van Buren, Jr., C. E. Illustrated, 8vo, cloth. 2 00

JOYNSON. Designing and Construction of Machine Gearing. Illustrated, 8vo, cloth..................... 2 00

GILLMORE. Coignet Beton and other Artificial Stone. By Q. A. Gillmore, Major U. S. Corps Engineers. 9 plates, views, &c. 8vo, cloth................... 2 50

SAELTZER. Treattse on Acoustics in connection with Ventilation. By Alexander Saeltzer, Architect. 12mo, cloth.. 2 00

THE EARTH'S CRUST. A handy Outline of Geology. By David Page. Illustrated, 18mo, cloth.... 75

DICTIONARY of Manufactures, Mining, Machinery, and the Industrial Arts. By George Dodd. 12mo, cloth.. 2 00

FRANCIS. On the Strength of Cast-Iron Pillars, with Tables for the use of Engineers, Architects, and Builders. By James B. Francis, Civil Engineer. 1 vol. 8vo, cloth.................................... 2 00

GILLMORE (Gen. Q. A.) Treatise on Limes, Hydraulic Cements, and Mortars. Papers on Practical Engineering, U. S. Engineer Department, No. 9, containing Reports of numerous Experiments conducted in New York City, during the years 1858 to 1861, inclusive. By Q. A. Gillmore, Bvt. Maj.-Gen., U. S. A., Major, Corps of Engineers. With numerous illustrations. 1 vol, 8vo, cloth............... $4 00

HARRISON. The Mechanic's Tool Book, with Practical Rules and Suggestions for Use of Machinists, Iron Workers, and others. By W. B. Harrison, associate editor of the "American Artisan." Illustrated with 44 engravings. 12mo, cloth............ 1 50

HENRICI (Olaus). Skeleton Structures, especially in their application to the Building of Steel and Iron Bridges. By Olaus Henrici. With folding plates and diagrams. 1 vol. 8vo, cloth.................. 3 00

HEWSON (Wm.) Principles and Practice of Embanking Lands from River Floods, as applied to the Levees of the Mississippi. By William Hewson, Civil Engineer. 1 vol. 8vo, cloth...................... 2 00

HOLLEY (A. L.) Railway Practice. American and European Railway Practice, in the economical Generation of Steam, including the Materials and Construction of Coal-burning Boilers, Combustion, the Variable Blast, Vaporization, Circulation, Superheating, Supplying and Heating Feed-water, etc., and the Adaptation of Wood and Coke-burning Engines to Coal-burning; and in Permanent Way, including Road-bed, Sleepers, Rails, Joint-fastenings, Street Railways, etc., etc. By Alexander L. Holley, B. P. With 77 lithographed plates. 1 vol. folio, cloth.... 12 00

KING (W. H.) Lessons and Practical Notes on Steam, the Steam Engine, Propellers, etc., etc., for Young Marine Engineers, Students, and others. By the late W. H. King, U. S. Navy. Revised by Chief Engineer J. W. King, U. S. Navy. Twelfth edition, enlarged. 8vo, cloth. 2 00

MINIFIE (Wm.) Mechanical Drawing. A Text-Book of Geometrical Drawing for the use of Mechanics and Schools, in which the Definitions and Rules of Geometry are familiarly explained; the Practical Problems are arranged, from the most simple to the more complex, and in their description technicalities are avoided as much as possible. With illustrations for Drawing Plans, Sections, and Elevations of Railways and Machinery; an Introduction to Isometrical Drawing, and an Essay on Linear Perspective and Shadows. Illustrated with over 200 diagrams engraved on steel. By Wm. Minifie, Architect. Seventh edition. With an Appendix on the Theory and Application of Colors. 1 vol. 8vo, cloth........... $4 00

"It is the best work on Drawing that we have ever seen, and is especially a text-book of Geometrical Drawing for the use of Mechanics and Schools. No young Mechanic, such as a Machinists, Engineer, Cabinet-maker, Millwright, or Carpenter, should be without it."—*Scientific American.*

——— Geometrical Drawing. Abridged from the octavo edition, for the use of Schools. Illustrated with 48 steel plates. Fifth edition. 1 vol. 12mo, cloth.... 2 00

STILLMAN (Paul.) Steam Engine Indicator, and the Improved Manometer Steam and Vacuum Gauges— their Utility and Application. By Paul Stillman. New edition. 1 vol. 12mo, flexible cloth........... 1 00

SWEET (S. H.) Special Report on Coal; showing its Distribution, Classification, and cost delivered over different routes to various points in the State of New York, and the principal cities on the Atlantic Coast. By S. H. Sweet. With maps, 1 vol. 8vo, cloth..... 3 00

WALKER (W. H.) Screw Propulsion. Notes on Screw Propulsion: its Rise and History. By Capt. W. H. Walker, U. S. Navy. 1 vol. 8vo, cloth..... 75

WARD (J. H.) Steam for the Million. A popular Treatise on Steam and its Application to the Useful Arts, especially to Navigation. By J. H. Ward, Commander U. S. Navy. New and revised edition. 1 vol. 8vo, cloth................................ 1 00

WEISBACH (Julius). Principles of the Mechanics of Machinery and Engineering. By Dr. Julius Weisbach, of Freiburg. Translated from the last German edition. Vol. I., 8vo, cloth.. .... 10 00

DIEDRICH. The Theory of Strains, a Compendium for the calculation and construction of Bridges, Roofs, and Cranes, with the application of Trigonometrical Notes, containing the most comprehensive information in regard to the Resulting strains for a permanent Load, as also for a combined (Permanent and Rolling) Load. In two sections, adadted to the requirements of the present time. By John Diedrich, C. E. Illustrated by numerous plates and diagrams. 8vo, cloth.. 5 00

WILLIAMSON (R. S.) On the use of the Barometer on Surveys and Reconnoissances. Part I. Meteorology in its Connection with Hypsometry. Part II. Barometric Hypsometry. By R. S. Wiliamson, Bvt. Lieut.-Col. U. S. A., Major Corps of Engineers. With Illustrative Tables and Engravings. Paper No. 15, Professional Papers, Corps of Engineers. 1 vol. 4to, cloth.. 15 00

POOK (S. M.) Method of Comparing the Lines and Draughting Vessels Propelled by Sail or Steam. Including a chapter on Laying off on the Mould-Loft Floor. By Samuel M. Pook, Naval Constructor. 1 vol. 8vo, with illustrations, cloth............ 5 00

ALEXANDER (J. H.) Universal Dictionary of Weights and Measures, Ancient and Modern, reduced to the standards of the United States of America. By J. H. Alexander. New edition, enlarged. 1 vol. 8vo, cloth.. 3 50

BROOKLYN WATER WORKS. Containing a Descriptive Account of the Construction of the Works, and also Reports on the Brooklyn, Hartford, Belleville and Cambridge Pumping Engines. With illustrations. 1 vol. folio, cloth................................

RICHARDS' INDICATOR. A Treatise on the Richards Steam Engine Indicator, with an Appendix by F. W. Bacon, M. E. 18mo, flexible, cloth.......... 1 00

POPE. Modern Practice of the Electric Telegraph. A Hand Book for Electricians and operators. By Frank L. Pope. Eighth edition, revised and enlarged, and fully illustrated. 8vo, cloth...................... $2.00

"There is no other work of this kind in the English language that contains in so small a compass so much practical information in the application of galvanic electricity to telegraphy. It should be in the hands of every one interested in telegraphy, or the use of Batteries for other purposes."

MORSE. Examination of the Telegraphic Apparatus and the Processes in Telegraphy. By Samuel F. Morse, LL.D., U. S. Commissioner Paris Universal Exposition, 1867. Illustrated, 8vo, cloth.......... $2 00

SABINE. History and Progress of the Electric Telegraph, with descriptions of some of the apparatus. By Robert Sabine, C. E. Second edition, with additions, 12mo, cloth.............................. 1 25

CULLEY. A Hand-Book of Practical Telegraphy. By R. S. Culley, Engineer to the Electric and International Telegraph Company. Fourth edition, revised and enlarged. Illustrated 8vo, cloth.............. 5 00

BENET. Electro-Ballistic Machines, and the Schultz Chronoscope. By Lieut.-Col. S. V. Benet, Captain of Ordnance, U. S. Army. Illustrated, second edition, 4to, cloth................................... 3 00

MICHAELIS. The Le Boulenge Chronograph, with three Lithograph folding plates of illustrations. By Brevet Captain O. E. Michaelis, First Lieutenant Ordnance Corps, U. S. Army, 4to, cloth.......... 3 00

ENGINEERING FACTS AND FIGURES An Annual Register of Progress in Mechanical Engineering and Construction, for the years 1863, 64, 65, 66, 67, 68. Fully illustrated, 6 vols. 18mo, cloth, $2.50 per vol., each volume sold separately..............

HAMILTON. Useful Information for Railway Men. Compiled by W. G. Hamilton, Engineer. Fifth edition, revised and enlarged, 562 pages Pocket form. Morocco, gilt.................................. 2 00

STUART. The Civil and Military Engineers of America. By Gen. C. B. Stuart. With 9 finely executed portraits of eminent engineers, and illustrated by engravings of some of the most important works constructed in America. 8vo, cloth.................... $5 00

STONEY. The Theory of Strains in Girders and similar structures, with observations on the application of Theory to Practice, and Tables of Strength and other properties of Materials. By Bindon B. Stoney, B. A. New and revised edition, enlarged, with numerous engravings on wood, by Oldham. Royal 8vo, 664 pages. Complete in one volume. 8vo, cloth....... 15 00

SHREVE. A Treatise on the Strength of Bridges and Roofs. Comprising the determination of Algebraic formulas for strains in Horizontal, Inclined or Rafter, Triangular, Bowstring, Lenticular and other Trusses, from fixed and moving loads, with practical applications and examples, for the use of Students and Engineers. By Samuel H. Shreve, A. M., Civil Engineer. 87 wood cut illustrations. 8vo, cloth.............. 5 00

MERRILL. Iron Truss Bridges for Railroads. The method of calculating strains in Trusses, with a careful comparison of the most prominent Trusses, in reference to economy in combination, etc., etc. By Brevet. Col. William E. Merrill, U. S. A., Major Corps of Engineers, with nine lithographed plates of Illustrations. 4to, cloth......................... 5 00

WHIPPLE. An Elementary and Practical Treatise on Bridge Building. An enlarged and improved edition of the author's original work. By S. Whipple, C. E., inventor of the Whipple Bridges, &c. Illustrated 8vo, cloth...................................... 4 00

THE KANSAS CITY BRIDGE. With an account of the Regimen of the Missouri River, and a description of the methods used for Founding in that River. By O. Chanute, Chief Engineer, and George Morrison, Assistant Engineer. Illustrated with five lithographic views and twelve plates of plans. 4to, cloth, 6 00

MAC CORD. A Practical Treatise on the Slide Valve
by Eccentrics, examining by methods the action of the
Eccentric upon the Slide Valve, and explaining the
Practical processes of laying out the movements,
adapting the valve for its various duties in the steam
engine. For the use of Engineers, Draughtsmen,
Machinists, and Students of Valve Motions in general. By C. W. Mac Cord, A. M., Professor of Mechanical Drawing, Stevens' Institute of Technology,
Hoboken, N. J. Illustrated by 8 full page copperplates. 4to, cloth................................ $4 00

KIRKWOOD. Report on the Filtration of River
Waters, for the supply of cities, as practised in
Europe, made to the Board of Water Commissioners
of the City of St. Louis. By James P. Kirkwood.
Illustrated by 30 double plate engravings. 4to, cloth, 15 00

PLATTNER. Manual of Qualitative and Quantitative
Analysis with the Blow Pipe. From the last German
edition, revised and enlarged. By Prof. Th. Richter.
of the Royal Saxon Mining Academy. Translated
by Prof. H. B. Cornwall, Assistant in the Columbia
School of Mines, New York, assisted by John H.
Caswell. Illustrated with 87 wood cuts, and one
lithographic plate. Second edition, revised, 560
pages, 8vo, cloth................................. 7 50

PLYMPTON. The Blow Pipe. A system of Instruc-
tion in its practical use being a graduated course of
analysis for the use of students, and all those engaged
in the examination of metallic combinations. Second
edition, with an appendix and a copious index. By
Prof. Geo. W. Plympton, of the Polytechnic Institute, Brooklyn, N. Y. 12mo, cloth................ 2 00

PYNCHON. Introduction to Chemical Physics, designed for the use of Academies, Colleges and High
Schools. Illustrated with numerous engravings, and
containing copious experiments with directions for
preparing them. By Thomas Ruggles Pynchon,
M. A., Professor of Chemistry and the Natural Sciences, Trinity College, Hartford. New edition, revised and enlarged. and illustrated by 269 illustrations
on wood. Crown, 8vo. cloth...................... 3 00

ELIOT AND STORER. A compendious Manual of Qualitative Chemical Analysis. By Charles W. Eliot and Frank H. Storer. Revised with the Co-operation of the authors. By William R. Nichols, Professor of Chemistry in the Massachusetts Institute of Technology. Illustrated, 12mo, cloth....... $1 50

RAMMELSBERG. Guide to a course of Quantitative Chemical Analysis, especially of Minerals and Furnace Products. Illustrated by Examples. By C. F. Rammelsberg. Translated by J. Towler, M. D. 8vo, cloth............................. 2 25

EGLESTON. Lectures on Descriptive Mineralogy, delivered at the School of Mines, Columbia College. By Professor T. Egleston. Illustrated by 34 Lithographic Plates. 8vo, cloth........................ 4 50

MITCHELL. A Manual of Practical Assaying. By John Mitchell. Third edition. Edited by William Crookes, F. R. S. 8vo, cloth................. 10 00

WATT'S Dictionary of Chemistry. New and Revised edition complete in 6 vols. 8vo. cloth, $62.00. Supplementary volume sold separately. Price, cloth... 9 00

RANDALL. Quartz Operators Hand-Book. By P. M. Randall. New edition, revised and enlarged, fully illustrated. 12mo, cloth...................... 2 00

SILVERSMITH. A Practical Hand-Book for Miners, Metallurgists, and Assayers, comprising the most recent improvements in the disintegration, amalgamation, smelting, and parting of the Precious ores, with a comprehensive Digest of the Mining Laws. Greatly augmented, revised and corrected. By Julius Silversmith. Fourth edition. Profusely illustrated. 12mo, cloth... 3 00

THE USEFUL METALS AND THEIR ALLOYS, including Mining Ventilation, Mining Jurisprudence, and Metallurgic Chemistry employed in the conversion of Iron, Copper, Tin, Zinc, Antimony and Lead ores, with their applications to the Industrial Arts. By Scoffren, Truan, Clay, Oxland, Fairbairn, and others. Fifth edition, half calf..................... 3 75

JOYNSON. The Metals used in construction, Iron, Steel, Bessemer Metal, etc., etc. By F. H. Joynson. Illustrated, 12mo, cloth.......................... $0 75

VON COTTA. Treatise on Ore Deposits. By Bernhard Von Cotta, Professor of Geology in the Royal School of Mines, Freidberg, Saxony. Translated from the second German edition, by Frederick Prime, Jr., Mining Engineer, and revised by the author, with numerous illustrations. 8vo, cloth........ 4 00

URE. Dictionary of Arts, Manufactures and Mines. By Andrew Ure, M.D. Sixth edition, edited by Robert Hunt, F. R. S., greatly enlarged and re-written. London, 1872. 3 vols. 8vo, cloth, $25.00. Half Russia.................................... 37 50

BELL. Chemical Phenomena of Iron Smelting. An experimental and practical examination of the circumstances which determine the capacity of the Blast Furnace, The Temperature of the air, and the proper condition of the Materials to be operated upon. By I. Lowthian Bell. 8vo, cloth........... 6 00

ROGERS. The Geology of Pennsylvania. A Government survey, with a general view of the Geology of the United States, Essays on the Coal Formation and its Fossils, and a description of the Coal Fields of North America and Great Britain. By Henry Darwin Rogers, late State Geologist of Pennsylvania, Splendidly illustrated with Plates and Engravings in the text. 3 vols., 4to, cloth, with Portfolio of Maps. 30 00

BURGH. Modern Marine Engineering, applied to Paddle and Screw Propulsion. Consisting of 36 colored plates, 259 Practical Wood Cut Illustrations, and 403 pages of descriptive matter, the whole being an exposition of the present practice of James Watt & Co., J. & G. Rennie, R. Napier & Sons, and other celebrated firms, by N. P. Burgh, Engineer, thick 4to, vol., cloth, $25.00 ; half mor........ 30 00

BARTOL. Treatise on the Marine Boilers of the United States. By B. H. Bartol. Illustrated, 8vo, cloth... 1 50

11

BOURNE. Treatise on the Steam Engine in its various applications to Mines, Mills, Steam Navigation, Railways, and Agriculture, with the theoretical investigations respecting the Motive Power of Heat, and the proper proportions of steam engines. Elaborate tables of the right dimensions of every part, and Practical Instructions for the manufacture and management of every species of Engine in actual use. By John Bourne, being the ninth edition of "A Treatise on the Steam Engine," by the "Artizan Club." Illustrated by 38 plates and 546 wood cuts. 4to, cloth..$15 00

STUART. The Naval Dry Docks of the United States. By Charles B. Stuart late Engineer-in-Chief of the U. S. Navy. Illustrated with 24 engravings on steel. Fourth edition, cloth..................... 6 00

EADS. System of Naval Defences. By James B. Eads, C. E., with 10 illustrations, 4to, cloth........ 5 00

FOSTER. Submarine Blasting in Boston Harbor, Massachusetts. Removal of Tower and Corwin Rocks. By J. G. Foster, Lieut.-Col. of Engineers, U. S. Army. Illustrated with seven plates, 4to, cloth... 3 50

BARNES Submarine Warfare, offensive and defensive, including a discussion of the offensive Torpedo System, its effects upon Iron Clad Ship Systems and influence upon future naval wars. By Lieut.-Commander J. S. Barnes, U. S. N., with twenty lithographic plates and many wood cuts. 8vo, cloth..... 5 00

HOLLEY. A Treatise on Ordnance and Armor, embracing descriptions, discussions, and professional opinions concerning the materials, fabrication, requirements, capabilities, and endurance of European and American Guns, for Naval, Sea Coast, and Iron Clad Warfare, and their Rifling, Projectiles, and Breech-Loading ; also, results of experiments against armor, from official records, with an appendix referring to Gun Cotton, Hooped Guns, etc., etc. By Alexander L. Holley, B. P., 948 pages, 493 engravings, and 147 Tables of Results, etc., 8vo, half roan. 10 00

12

SIMMS. A Treatise on the Principles and Practice of Levelling, showing its application to purposes of Railway Engineering and the Construction of Roads, &c. By Frederick W. Simms, C. E. From the 5th London edition, revised and corrected, with the addition of Mr. Laws's Practical Examples for setting out Railway Curves. Illustrated with three Lithographic plates and numerous wood cuts. 8vo, cloth. **$2 50**

BURT. Key to the Solar Compass, and Surveyor's Companion ; comprising all the rules necessary for use in the field; also description of the Linear Surveys and Public Land System of the United States, Notes on the Barometer, suggestions for an outfit for a survey of four months, etc. By W. A. Burt, U. S. Deputy Surveyor. Second edition. Pocket book form, tuck... 2 50

THE PLANE TABLE. Its uses in Topographical Surveying, from the Papers of the U. S. Coast Survey. Illustrated, 8vo, cloth...................... 2 00

"This work gives a description of the Plane Table, employed at the U. S. Coast Survey office, and the manner of using it."

JEFFER'S. Nautical Surveying: By W. N. Jeffers, Captain U. S. Navy. Illustrated with 9 copperplates and 31 wood cut illustrations. 8vo, cloth.......... 5 00

CHAUVENET. New method of correcting Lunar Distances, and improved method of Finding the error and rate of a chronometer, by equal altitudes. By W. Chauvenet, LL. D. 8vo, cloth................ 2 00

BRUNNOW. Spherical Astronomy. By F. Brunnow, Ph. Dr. Translated by the author from the second German edition. 8vo, cloth.................... 6 50

PEIRCE. System of Analytic Mechanics. By Benjamin Peirce. 4to, cloth........................ 10 00

COFFIN. Navigation and Nautical Astronomy. Prepared for the use of the U. S. Naval Academy. By Prof. J. H. C. Coffin. Fifth edition. 52 wood cut illustrations. 12mo, cloth 3 50

13

CLARK. Theoretical Navigation and Nautical Astronomy. By Lieut. Lewis Clark, U. S. N. Illustrated with 41 wood cuts. 8vo, cloth...................... **$3** 00

HASKINS. The Galvanometer and its Uses. A Manual for Electricians and Students. By C. H. Haskins. 12mo, pocket form, morocco. (In press).....

GOUGE. New System of Ventilation, which has been thoroughly tested, under the patronage of many distinguished persons. By Henry A. Gouge. With many illustrations. 8vo, cloth...................... 2 00

BECKWITH. Observations on the Materials and Manufacture of Terra-Cotta, Stone Ware, Fire Brick, Porcelain and Encaustic Tiles, with remarks on the products exhibited at the London International Exhibition, 1871. By Arthur Beckwith, C. E. 8vo, paper... 60

MORFIT. A Practical Treatise on Pure Fertilizers, and the chemical conversion of Rock Guano, Marlstones, Coprolites, and the Crude Phosphates of Lime and Alumina generally, into various valuable products. By Campbell Morfit, M.D., with 28 illustrative plates, 8vo, cloth.................................. .... 20 00

BARNARD. The Metric System of Weights and Measures. An address delivered before the convocation of the University of the State of New York, at Albany, August, 1871. By F. A. P. Barnard, LL.D., President of Columbia College, New York. Second edition from the revised edition, printed for the Trustees of Columbia College. Tinted paper, 8vo, cloth 3 00

—— Report on Machinery and Processes on the Industrial Arts and Apparatus of the Exact Sciences. By F. A. P. Barnard, LL.D. Paris Universal Exposition, 1867. Illustrated, 8vo, cloth............. 5 00

BARLOW. Tables of Squares, Cubes, Square Roots, Cube Roots, Reciprocals of all integer numbers up to 10,000. New edition, 12mo, cloth................ 2 50

14

MYER. Manual of Signals, for the use of Signal officers in the Field, and for Military and Naval Students, Military Schools, etc. A new edition enlarged and illustrated. By Brig. General Albert J. Myer, Chief Signal Officer of the army, Colonel of the Signal Corps during the War of the Rebellion. 12mo, 48 plates, full Roan............................... $5 00

WILLIAMSON. Practical Tables in Meteorology and Hypsometry, in connection with the use of the Barometer. By Col. R. S. Williamson, U. S. A. 4to, cloth.. 2 50

THE YOUNG MECHANIC. Containing directions for the use of all kinds of tools, and for the construction of Steam Engines and Mechanical Models, including the Art of Turning in Wood and Metal. By the author "The Lathe and its Uses," etc. From the English edition with corrections. Illustrated, 12mo, cloth...................................... 1 75

PICKERT AND METCALF. The Art of Graining. How Acquired and How Produced, with description of colors, and their application. By Charles Pickert and Abraham Metcalf. Beautifully illustrated with 42 tinted plates of the various woods used in interior finishing. Tinted paper, 4to, cloth................ 10 00

HUNT. Designs for the Gateways of the Southern Entrances to the Central Park. By Richard M. Hunt. With a description of the designs. 4to. cloth...... 5 00

LAZELLE. One Law in Nature. By Capt. H. M. Lazelle, U. S. A. A new Corpuscular Theory, comprehending Unity of Force, Identity of Matter, and its Multiple Atom Constitution, applied to the Physical Affections or Modes of Energy. 12mo, cloth... 1 50

PETERS. Notes on the Origin, Nature, Prevention, and Treatment of Asiatic Cholera. By John C. Peters, M. D. Second edition, with an Appendix. 12mo, cloth.. 1 50

BOYNTON. History of West Point, its Military Importance during the American Revolution, and the Origin and History of the U. S. Military Academy. By Bvt. Major C. E. Boynton, A.M., Adjutant of the Military Academy. Second edition, 416 pp. 8vo, printed on tinted paper, beautifully illustrated with 36 maps and fine engravings, chiefly from photographs taken on the spot by the author. Extra cloth... $3 50

WOOD. West Point Scrap Book, being a collection of Legends, Stories, Songs, etc., of the U. S. Military Academy. By Lieut. O. E. Wood, U. S. A. Illustrated by 69 engravings and a copperplate map. Beautifully printed on tinted paper. 8vo, cloth..... 5 00

WEST POINT LIFE. A Poem read before the Dialectic Society of the United States Military Academy. Illustrated with Pen-and-Ink Sketches. By a Cadet. To which is added the song, "Benny Havens, oh !" oblong 8vo, 21 full page illustrations, cloth.......... 2 50

GUIDE TO WEST POINT and the U. S. Military Academy, with maps and engravings, 18mo, blue cloth, flexible.................................... 1 00

HENRY. Military Record of Civilian Appointments in the United States Army. By Guy V. Henry, Brevet Colonel and Captain First United States Artillery, Late Colonel and Brevet Brigadier General, United States Volunteers. Vol. 1 now ready. Vol. 2 in press. 8vo, per volume, cloth.................... 5 00

HAMERSLY. Records of Living Officers of the U. S. Navy and Marine Corps. Compiled from official sources. By Lewis B. Hamersly, late Lieutenant U. S. Marine Corps. Revised edition, 8vo, cloth... 5 00

MOORE. Portrait Gallery of the War. Civil, Military and Naval. A Biographical record, edited by Frank Moore. 60 fine portraits on steel. Royal 8vo, cloth... 6 00

www.ingramcontent.com/pod-product-compliance
Lightning Source LLC
Chambersburg PA
CBHW022005190326
41519CB00010B/1387